I0445371

First Printing: August 2022

TABLE OF CONTENTS

RECENT GED® TEST HISTORY

The GED® Math Reasoning Test was created in 1942 and continues to help students in America and around the world to certify their high school equivalency.

First, some basic facts about the producers of the GED® Test. In March of 2011, The American Council of Education (ACE), which was the only nonprofit "higher education organization that represents presidents and chancellors of all types of U.S. accredited, degree-granting institutions: community colleges, four-year institutions, private and public universities, and nonprofit and for-profit colleges" joined together with Pearson Vue, a for-profit corporation specializing in testing and education products, to advance GED® Testing. (http://www.acenet.edu)

The new GED Testing Service® (GEDTS) produced a new, enhanced version of the GED® Test in 2014. The contents of this book specifically address the current GED® Math Test (version as of March 1, 2016).

In 2013, the final year of the old 2002 version of the test, some 560,000 (1) students passed the GED® Test. In the first year of the new exam, 2014, only about 58,500 (2) passed the test. This was a dramatic drop! 2015 results were not much better. By January of 2016, the GEDTS was lowering its passing standard from 150 to 145 and removing the Extended Response (the essay) from the Social Studies test.

In order to pass the test you need to score at least 145 (in most US States—as of the printing of this book, some States have not yet approved this new passing score) out of 200 on each of the four tests (3).
Math—46 Questions, 115 Minutes
Science—34 Questions, 90 Minutes
Social Studies—35 Questions, 70 Minutes
RLA—46 Questions, 150 Minutes

GOOD NEWS!!!
The GEDTS® actually publishes the specific skills and knowledge being tested on each test and what percent of those skills make up the test.

Therefore, here is my approach:
to eliminate the most difficult areas of knowledge and skills (30-35% of the whole);
to organize and prepare for study 65-70% of the knowledge areas and skills tested

My approach allows you to study only enough to pass the test. It works. I have over 12 years and thousands of successful students to prove it!

You can miss over 30% of the knowledge and skills being tested and still pass the test! With the new passing score standard of 145, my method is even more successful!

As I say often, for most, the GED® is not about passing with high honors and high scores. Either you pass or you fail. The Get Your GED® Now Test Preparation Series is focused on you passing, not scoring high. You've got the rest of your life to score high in matters that interest you. Just pass the GED® Test and get on with the rest of your life...that's my motto!

However, I do need to acknowledge that beyond passing the GED® Test, the GEDTS offers the following:
145—164—Pass / High School Equivalency
165—174—GED® College Ready
175—200—GED® College Ready + Credit

Damon A. Tinnon
Creator of the Get Your GED Now Test Preparation Series

1. The 2013 Annual Statistical Report on the GED® Test.
2. "A 'Sizable Decrease' In Those Passing The GED." January 9, 2015. www.npr.org.
3. GED.COM.

GED® MATH TEST

If you want to pass the GED® Math Test, you need to start with the basic facts of the test. The GED® Math Test is organized into two parts:
Part 1 is 5 questions, no calculator allowed
Part 2 is 41 questions, calculator is allowed

The test length is 115 minutes.

The GED® Math Test has two main subject areas:
45% is basic math
55% is Algebra

Remember, this test is designed and written by the GEDTS®. Below is a screenshot from their website that gives even more detail of what you can expect to see on the test.
Of the 45% of the GED® Math Test that is Basic Math, 25% is:

Mathematical Reasoning Skill Areas	Question Numbers
Mathematical Reasoning Skill Area 1: Quantitative problem solving with rational numbers (25% of the 2014 GED® Mathematical Reasoning test)	
Place fractions and decimals in order, including when using a number line	
Apply number properties that involve multiples and factors	
Simplify numerical expressions with rational exponents	
Identify the absolute value of a rational number as its distance from 0 on the number line and find the distance between two rational numbers on the number line	
Compute with rational numbers and solve problems with rational numbers	
Write and compute with numerical expressions with squares, square roots, cubes, and cube roots of positive, rational numbers	
Determine when a numerical expression is undefined	
Compute unit rates	1
Use scale factors to figure out the magnitude of a size change and convert between actual drawings and scale drawings	
Solve two-step, arithmetic, real-world problems that involve ratios, proportions, and percents	4

GED® MATH TEST

Another screenshot from GEDTS® is below. Of the 45% of the GED® Math Test that is Basic Math, 20% is:

Mathematical Reasoning Skill Area 2: Quantitative problem solving in measurement (20% of the 2014 GED® Mathematical Reasoning test)	
Compute the area and perimeter of various shapes: triangles, rectangles, polygons, and composite figures	
Find the side lengths of triangles, rectangles, and polygons when given the area or perimeter	
Compute the area and circumference of circles and find the radius or diameter when given the area or circumference	
Use the Pythagorean theorem ($a^2 + b^2 = c^2$) to determine unknown side lengths in a right triangle	
Compute volume and surface area of right prisms and pyramids, cylinders, spheres, cones, and composite figures	7, 9
Solve for height, radius, diameter, or side lengths of cylinders, cones, and right pyramids, when given the volume or surface area	
Represent, display, and interpret categorical data in bar graphs, circle graphs, dot plots, histograms, box plots, tables, and scatter plots	
Calculate the mean, median, mode, range, and weighted average, and calculate a missing data value, given the average and all the missing data values but one	
Use counting techniques to solve problems and find combinations and permutations	
Find the probability of simple and compound events	

> **POWER POINT**
> The key to passing the GED® Math Test is studying the right material! Don't waste your time studying things that won't help you!

The Calculator

You can bring a TI-30XS Multiview Scientific Calculator to use on the GED® test or an on-screen TI-30XS Multiview Scientific Calculator is provided.

1. Calculator Picture Source: GED.COM.

GED® MATH TEST

The Formulas

The formulas necessary for the test are provided and there is no need to memorize them. Below is a screenshot of these formulas.

Mathematics Formula Sheet

Area of a:

square	$A = s^2$
rectangle	$A = lw$
parallelogram	$A = bh$
triangle	$A = \frac{1}{2}bh$
trapezoid	$A = \frac{1}{2}h(b_1 + b_2)$
circle	$A = \pi r^2$

Perimeter of a:

square	$P = 4s$
rectangle	$P = 2l + 2w$
triangle	$P = s_1 + s_2 + s_3$
Circumference of a circle	$C = 2\pi r$ OR $C = \pi d$; $\pi \approx 3.14$

Surface area and volume of a:

rectangular prism	$SA = 2lw + 2lh + 2wh$	$V = lwh$
right prism	$SA = ph + 2B$	$V = Bh$
cylinder	$SA = 2\pi rh + 2\pi r^2$	$V = \pi r^2 h$
pyramid	$SA = \frac{1}{2}ps + B$	$V = \frac{1}{3}Bh$
cone	$SA = \pi rs + \pi r^2$	$V = \frac{1}{3}\pi r^2 h$
sphere	$SA = 4\pi r^2$	$V = \frac{4}{3}\pi r^3$

(p = perimeter of base with area B, $\pi \approx 3.14$)

Data

mean	mean is equal to the total of the values of a data set, divided by the number of elements in the data set
median	median is the middle value in an odd number of ordered values of a data set, or the mean of the two middle values in an even number of ordered values in a data set

Algebra

slope of a line	$m = \dfrac{y_2 - y_1}{x_2 - x_1}$
slope-intercept form of the equation of a line	$y = mx + b$
point-slope form of the equation of a line	$y - y_1 = m(x - x_1)$
standard form of a quadratic equation	$y = ax^2 + bx + c$
quadratic formula	$x = \dfrac{-b \pm \sqrt{b^2 - 4ac}}{2a}$
Pythagorean theorem	$a^2 + b^2 = c^2$
simple interest	$I = Prt$ (I = interest, P = principal, r = rate, t = time)
distance formula	$d = rt$
total cost	total cost = (number of units) × (price per unit)

1. Formula Page Source: GED.COM.

The Symbols

The symbols necessary for the test are provided and there is no need to memorize them. Below is a screenshot of the symbols:

Æ Symbol Tool Explanation

The 2014 GED® test on computer contains a tool known as the "Æ Symbol Tool." Use this guide to learn about entering special mathematical symbols into fill-in-the-blank item types.

Symbol	Explanation	Symbol	Explanation	Symbol	Explanation
Π	pi	|	absolute value	–	minus or negative
f	function	×	multiplication	(	open or left parenthesis
$\geq$	greater than or equal to	÷	division	)	close or right parenthesis
$\leq$	less than or equal to	±	positive or negative	>	greater than
$\neq$	not equal to	∞	infinity	<	less than
2	2 exponent ("squared")	$\sqrt{}$	square root	=	equals
3	3 exponent ("cubed")	+	plus or positive		

POWER POINT
To pass the test you need to score 145 in most states* out of a possible 200 score. If you can get 65%-70% of the questions correct, you can pass the GED Math Test!

*In January 2016, the GEDTS® changed the passing standard of the GED® Test from 150 to 145 for all tests. As of the printing of this workbook, all US States that currently recognize the GED® Test endorsed and accepted this change.

1. Symbol Page Source: GED.COM.

GED® MATH STRATEGY

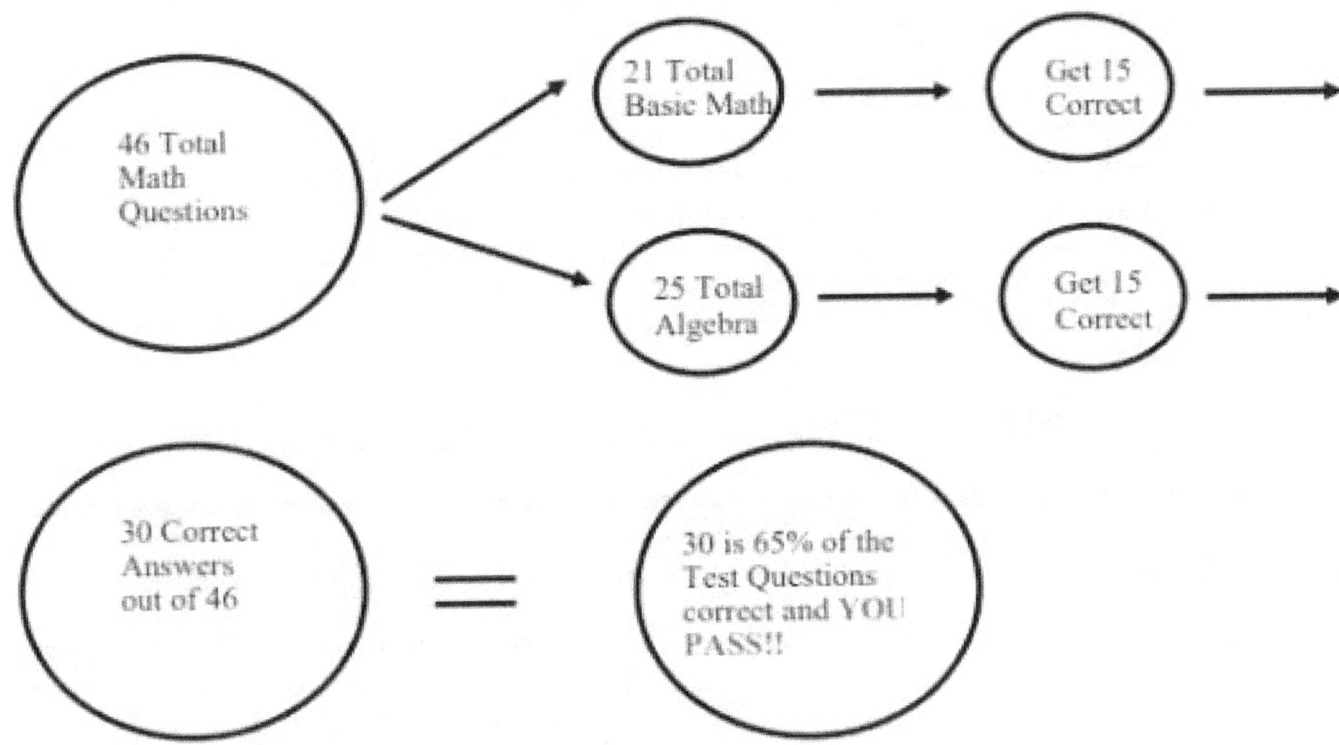

My strategy to help you pass the math test is simple. There are 46 questions on the math test. 45% of those questions are basic math questions which equals about 21 questions. 55% of those questions are Algebra questions which equals about 25 questions.

Of these 46 questions, you need to get about 65% or about 30 correct to get the 145 passing standard.

If you can get most of the basic math questions correct—15 correct out of 21 and if you can get at least 15 of the Algebra questions correct—15 out of 25—you will pass the test!

So, I simply use the design of the test itself to help you pass the test. This allows me to lay before you a clear and simple path and cut out the most difficult material!

This workbook will walk you through the basic math questions you will need to know to get 15 of the basic math questions correct and the algebraic math questions you will need to know to get 15 of the algebraic math questions correct.

*Note—not every question has the same point value, but in general, you can be confident that getting 65% - 70% of the test questions correct, across a representative spread of the question types, will provide you enough points to earn a passing score.

Adding Fractions

The GED® Math Reasoning Test presents problems involving the adding of fractions.

The focus of this lesson is to learn to add fractions. The concept is used throughout basic math and algebra. Fractions are tested as a skill and also as word problems. The key is to see the connection between the skill and the word problem. Word problems become easier to solve when you can find the simple math within the text!

ADDING FRACTIONS - EXAMPLE

SKILL EXAMPLE

Find the sum.

1.
$$+ \frac{\frac{2}{5}}{\frac{1}{5}}$$
_____

WORD PROBLEM EXAMPLE

Nila needed to interview 35 applicant. On Monday her interviewed 2/5 of the applicants. On Tuesday, she interviewed 1 / 5 of the applicants. What fraction of the applicants had Nila interviewed by Tuesday?

INDENTIFY THE QUESTION
First, identify that the question asks what fraction of the applicants had been interviewed

STATE YOUR PATH
>Interviewed 2/5 on Monday
>Interviewed 1/5 on Tuesday
>2/5 + 1/5 = 3/5 of the applicants had been interviewed

Video Explanation

To view video explanation of example, go to: https://ispri.ng/JkD31

ADDING FRACTIONS - SKILL PROBLEMS

Fractions Addition

Find the sum.

2. $\frac{7}{8}$ $+ \frac{4}{8}$ _____

3. $\frac{4}{6}$ $+ \frac{4}{6}$ _____

4. $\frac{5}{8}$ $+ \frac{1}{8}$ _____

5. $\frac{1}{5}$ $+ \frac{3}{5}$ _____

6. $\frac{2}{4}$ $+ \frac{2}{4}$ _____

7. $\frac{1}{3}$ $+ \frac{2}{3}$ _____

8. $\frac{1}{8}$ $+ \frac{3}{8}$ _____

9. $\frac{1}{6}$ $+ \frac{4}{6}$ _____

10. $\frac{4}{5}$ $+ \frac{2}{5}$ _____

11. $\frac{3}{4}$ $+ \frac{3}{4}$ _____

12. $\frac{3}{4}$ $+ \frac{1}{4}$ _____

13. $\frac{1}{3}$ $+ \frac{1}{3}$ _____

14. $\frac{7}{8}$ $+ \frac{6}{8}$ _____

15. $\frac{4}{5}$ $+ \frac{4}{5}$ _____

16. $\frac{5}{6}$ $+ \frac{2}{6}$ _____

17. $\frac{2}{3}$ $+ \frac{1}{3}$ _____

18. $\frac{1}{4}$ $+ \frac{3}{4}$ _____

19. $\frac{3}{6}$ $+ \frac{2}{6}$ _____

20. $\frac{2}{3}$ $+ \frac{2}{3}$ _____

21. $\frac{1}{6}$ $+ \frac{2}{6}$ _____

22. $\frac{3}{5}$ $+ \frac{4}{5}$ _____

Mixed Operations with Fractions

Calculate.

23. $5\frac{5}{6} + 2\frac{1}{3} =$ ___

24. $7\frac{2}{4} + 2\frac{3}{5} =$ ___

25. $2\frac{2}{3} + 2\frac{4}{5} =$ ___

26. $1\frac{1}{6} + 1\frac{7}{8} =$ ___

27. $6\frac{2}{4} + 3\frac{1}{3} =$ ___

28. $6\frac{1}{4} + 3\frac{5}{8} =$ ___

29. $4\frac{2}{5} + 4\frac{4}{6} =$ ___

30. $3\frac{3}{6} + 4\frac{1}{3} =$ ___

31. $1\frac{3}{5} + 7\frac{3}{4} =$ ___

32. $5\frac{5}{8} + 3\frac{1}{6} =$ ___

33. $4\frac{2}{8} + 5\frac{2}{3} =$ ___

34. $1\frac{1}{4} + 2\frac{2}{5} =$ ___

35. $3\frac{4}{6} + 4\frac{3}{8} =$ ___

36. $9\frac{1}{5} + 6\frac{1}{4} =$ ___

37. $2\frac{2}{3} + 2\frac{2}{5} =$ ___

SECTION 1 ANSWERS

1. 3/5

2. 1 3/8	3. 1 1/3	4. 3/4	5. 4/5	6. 1	7. 1	8. 1/2	9. 5/6
10. 1 1/5	11. 1 1/2	12. 1	13. 2/3	14. 1 5/8	15. 1 3/5	16. 1 1/6	17. 1
18. 1	19. 5/6	20. 1 1/3	21. 1/2	22. 1 2/5			

23. 8 1/6	24. 10 1/10	25. 5 7/15	26. 3 1/24	27. 9 5/6	28. 9 7/8
29. 9 1/15	30. 7 5/6	31. 9 7/20	32. 8 19/24	33. 9 11/12	34. 3 13/20
35. 8 1/24	36. 15 9/20	37. 5 1/15			

ADDING FRACTION - WORD PROBLEMS

Word Problems - Fractions Addition

Solve.

38. A recipe calls for $\frac{2}{3}$ cups of white flour and $\frac{5}{8}$ cups of whole wheat flour. How much flour in total is needed for the recipe?

_____

39. Jackie cycled $\frac{1}{4}$ miles. She then stopped to have a snack. Then she cycled $\frac{1}{3}$ more miles. How far did Jackie cycle?

_____

40. What is the sum of $\frac{6}{8}$ and $\frac{3}{6}$?

_____

41. Billy ran $\frac{3}{5}$ of a mile and then walked another $\frac{3}{4}$ of a mile. How far did he travel?

_____

42. What is $\frac{3}{4}$ plus $\frac{2}{3}$?

_____

43. If the sum of two fractions is $1\frac{7}{15}$ and the first fraction is $\frac{2}{3}$, what is the second fraction?

_____

44. What is $\frac{1}{3}$ plus $\frac{1}{3}$?

_____

45. A recipe calls for $\frac{2}{5}$ cups of white flour and $\frac{2}{6}$ cups of whole wheat flour. How much flour in total is needed for the recipe?

_____

46. Sandra cycled $\frac{1}{4}$ miles. She then stopped to have a snack. Then she cycled $\frac{1}{8}$ more miles. How far did Sandra cycle?

_____

47. If the sum of two fractions is $\frac{11}{30}$ and the first fraction is $\frac{1}{5}$, what is the second fraction?

_____

48. What is the sum of $\frac{1}{3}$ and $\frac{4}{5}$?

_____

49. Steven ran $\frac{3}{8}$ of a mile and then walked another $\frac{1}{4}$ of a mile. How far did he travel?

_____

50. What is the sum of $\frac{2}{3}$ and $\frac{3}{5}$?

_____

51. Jake ran $\frac{1}{4}$ of a mile and then walked another $\frac{5}{6}$ of a mile. How far did he travel?

_____

52. A recipe calls for $\frac{2}{5}$ cups of white flour and $\frac{3}{4}$ cups of whole wheat flour. How much flour in total is needed for the recipe?

_____

SECTION 2 ANSWERS

38. 1 7/24   39. 7/12   40. 1 1/4   41. 1 7/20   42. 1 5/12   43. 4/5   44. 2/3

45. 11/15   46. 3/8   47. 1/6   48. 1 2/15   49. 5/8   50. 1 4/15   51. 1 1/12

52. 1 3/20

Subtracting Fractions

The GED® Math Reasoning Test presents problems involving the subtracting of fractions.

The focus of this lesson is to learn to subtract fractions. The concept is used throughout basic math and algebra. Fractions are tested as a skill and also as word problems. The key is to see the connection between the skill and the word problem. Word problems become easier to solve when you can find the simple math within the text!

SUBTRACTING FRACTIONS - EXAMPLE

SKILL EXAMPLE

Find the difference.

1.
$$\begin{array}{r} \frac{1}{2} \\ - \frac{1}{3} \\ \hline \end{array}$$

WORD PROBLEM EXAMPLE

Tasha had one-half of her project left to complete. If she completed 1/3 of the project on Monday. What fraction of the project would need to be completed on Tuesday?

INDENTIFY THE QUESTION
First, identify that the question asks what fraction would need to be completed on Tuesday

STATE YOUR PATH
>1/2 of project remaining
>1/3 completed on Monday
>1/2 - 1/3 = 1/6 of the project would need to be completed on Tuesday

Video Explanation

To view video explanation of example, go to: https://ispri.ng/JkD31

Fractions Subtraction

Find the difference.

2. $\frac{6}{8} - \frac{1}{4}$ 3. $\frac{3}{6} - \frac{1}{4}$ 4. $\frac{2}{5} - \frac{1}{5}$ 5. $\frac{3}{4} - \frac{1}{3}$ 6. $\frac{2}{4} - \frac{1}{3}$ 7. $\frac{5}{8} - \frac{2}{5}$ 8. $\frac{2}{3} - \frac{2}{5}$

9. $\frac{7}{8} - \frac{3}{4}$ 10. $\frac{3}{4} - \frac{1}{5}$ 11. $\frac{2}{4} - \frac{2}{6}$ 12. $\frac{5}{8} - \frac{1}{4}$ 13. $\frac{3}{8} - \frac{1}{8}$ 14. $\frac{4}{5} - \frac{2}{3}$ 15. $\frac{2}{3} - \frac{3}{6}$

16. $\frac{4}{6} - \frac{1}{4}$ 17. $\frac{2}{4} - \frac{2}{5}$ 18. $\frac{3}{8} - \frac{2}{8}$ 19. $\frac{4}{6} - \frac{5}{8}$ 20. $\frac{5}{8} - \frac{1}{3}$ 21. $\frac{7}{8} - \frac{1}{5}$ 22. $\frac{7}{8} - \frac{1}{6}$

Mixed Operations with Fractions

Calculate.

23. $9\frac{3}{4} - 1\frac{6}{8} =$ ___   24. $7\frac{4}{5} - 5\frac{2}{3} =$ ___   25. $2\frac{5}{6} - 1\frac{7}{8} =$ ___   26. $5\frac{1}{3} - 4\frac{3}{6} =$ ___   27. $7\frac{1}{5} - 6\frac{2}{4} =$ ___

28. $9\frac{2}{5} - 5\frac{7}{8} =$ ___   29. $9\frac{2}{3} - 8\frac{3}{6} =$ ___   30. $7\frac{1}{4} - 2\frac{2}{4} =$ ___   31. $7\frac{5}{8} - 6\frac{4}{5} =$ ___   32. $8\frac{2}{3} - 7\frac{1}{6} =$ ___

33. $8\frac{2}{5} - 7\frac{2}{3} =$ ___   34. $9\frac{1}{8} - 4\frac{2}{4} =$ ___   35. $8\frac{5}{6} - 6\frac{1}{3} =$ ___   36. $5\frac{3}{4} - 2\frac{4}{5} =$ ___   37. $8\frac{2}{6} - 6\frac{5}{8} =$ ___

SECTION 1 ANSWERS

1. 1/6

2. 1/2	3. 1/4	4. 1/5	5. 5/12	6. 1/6	7. 9/40	8. 4/15
9. 1/8	10. 11/20	11. 1/6	12. 3/8	13. 1/4	14. 2/15	15. 1/6
16. 5/12	17. 1/10	18. 1/8	19. 1/24	20. 7/24	21. 27/40	22. 17/24

23. 8   24. 2 2/15   25. 23/24   26. 5/6   27. 7/10   28. 3 21/40

29. 1 1/6   30. 4 3/4   31. 33/40   32. 1 1/2   33. 11/15   34. 4 5/8

35. 2 1/2   36. 2 19/20   37. 1 17/24

SUBTRACTING FRACTION - WORD PROBLEMS

Word Problems - Fractions Subtraction

Solve.

38. If you subtract $\frac{4}{8}$ from another fraction and the result is $\frac{3}{8}$, what was the other fraction?

_____

39. Ellen bought $\frac{2}{3}$ of a pound of raisins. After eating some of the raisins there was $\frac{1}{6}$ of a pound left over. How much did Ellen already eat?

_____

40. Marin walked $\frac{3}{6}$ of a mile on Monday and walked $\frac{2}{6}$ of a mile on Tuesday. How much further did Marin walk on Monday?

_____

41. If you subtract $\frac{1}{3}$ from $\frac{6}{8}$ what is the result?

_____

42. Allan is working on a project that requires a piece of wire that is $\frac{4}{6}$ of a meter long. He has a longer piece of wire that he cuts and removes $\frac{2}{15}$ of a meter to make it the right size. How long was the original piece of wire?

_____

43. David bought $\frac{2}{3}$ of a pound of jelly beans. He ate $\frac{1}{4}$ of a pound. How much was left?

_____

44. What is $\frac{4}{8}$ minus $\frac{2}{5}$?

_____

45. If you subtract $\frac{2}{8}$ from $\frac{2}{3}$ what is the result?

_____

46. What is $\frac{3}{4}$ minus $\frac{4}{6}$?

_____

47. Paul is working on a project that requires a piece of wire that is $\frac{3}{8}$ of a meter long. He has a longer piece of wire that he cuts and removes $\frac{9}{40}$ of a meter to make it the right size. How long was the original piece of wire?

_____

48. Michele bought $\frac{3}{4}$ of a pound of raisins. After eating some of the raisins there was $\frac{7}{20}$ of a pound left over. How much did Michele already eat?

_____

49. Adam bought $\frac{5}{6}$ of a pound of jelly beans. He ate $\frac{2}{4}$ of a pound. How much was left?

_____

50. Jake walked $\frac{4}{5}$ of a mile on Monday and walked $\frac{1}{5}$ of a mile on Tuesday. How much further did Jake walk on Monday?

_____

51. If you subtract $\frac{3}{5}$ from another fraction and the result is $\frac{1}{40}$, what was the other fraction?

_____

52. Jackie bought $\frac{3}{8}$ of a pound of raisins. After eating some of the raisins there was $\frac{1}{4}$ of a pound left over. How much did Jackie already eat?

_____

SECTION 2 ANSWERS

38. 7/8   39. 2/4   40. 1/6   41. 5/12   42. 4/5   43. 5/12   44. 1/10   45. 5/12

46. 1/12   47. 3/5   48. 2/5   49. 1/3   50. 3/5   51. 5/8   52. 1/8

Multiplying Fractions

The GED® Math Reasoning Test presents problems involving the multiplying of fractions.

The focus of this lesson is to learn to multiply fractions. The concept is used throughout basic math and algebra. Fractions are tested as a skill and also as word problems. The key is to see the connection between the skill and the word problem. Word problems become easier to solve when you can find the simple math within the text!

MULTIPLYING FRACTIONS - EXAMPLE

SKILL EXAMPLE

Find the product.

1. $\frac{1}{4} \times 24 =$ _____

WORD PROBLEM EXAMPLE

Mal received essays from his 24 students. If he had to read and grade 1/4 of the essays today, how many essays did Mal grade?

INDENTIFY THE QUESTION
First, identify that the question asks how many essays did Mal grade

STATE YOUR PATH
>Mal graded 1/4 of the essay
>Mal has 24 students
>1/4 X 24 = 6 essays graded

Video Explanation
To view video explanation of example, go to: https://ispri.ng/JkD31

MULTIPLYING FRACTIONS - SKILL PROBLEMS

Fractions Multiplication I

Find the product.

2. $\frac{3}{4} \times \frac{1}{6} =$ ___

3. $\frac{1}{6} \times \frac{4}{5} =$ ___

4. $\frac{3}{8} \times \frac{2}{3} =$ ___

5. $\frac{5}{8} \times \frac{1}{6} =$ ___

6. $\frac{2}{5} \times \frac{3}{5} =$ ___

7. $\frac{1}{4} \times \frac{2}{5} =$ ___

8. $\frac{6}{8} \times \frac{2}{3} =$ ___

9. $\frac{4}{6} \times \frac{1}{3} =$ ___

10. $\frac{3}{4} \times \frac{3}{4} =$ ___

11. $\frac{2}{5} \times \frac{1}{5} =$ ___

Fractions Multiplication II

Find the product.

12. $4 \times \frac{3}{5} =$ ___

13. $7 \times \frac{1}{3} =$ ___

14. $2 \times \frac{4}{6} =$ ___

15. $5 \times \frac{2}{3} =$ ___

16. $9 \times \frac{4}{8} =$ ___

17. $5 \times \frac{3}{4} =$ ___

18. $2 \times \frac{2}{3} =$ ___

19. $4 \times \frac{1}{4} =$ ___

20. $1 \times \frac{1}{6} =$ ___

21. $6 \times \frac{5}{8} =$ ___

Fractions Multiplication III

Find the product.

22. $1\frac{5}{8} \times \frac{4}{6} =$ ___

23. $9\frac{3}{4} \times \frac{3}{8} =$ ___

24. $2\frac{1}{3} \times \frac{3}{4} =$ ___

25. $2\frac{3}{4} \times \frac{1}{3} =$ ___

26. $8\frac{4}{8} \times \frac{3}{8} =$ ___

27. $9\frac{1}{4} \times \frac{2}{3} =$ ___

28. $6\frac{2}{6} \times \frac{1}{5} =$ ___

29. $6\frac{2}{5} \times \frac{4}{6} =$ ___

30. $9\frac{3}{4} \times \frac{1}{5} =$ ___

31. $6\frac{2}{3} \times \frac{3}{4} =$ ___

Fractions Multiplication IV

Find the product.

32. $5\frac{2}{4} \times 7\frac{2}{4} =$ ___

33. $6\frac{1}{5} \times 1\frac{4}{8} =$ ___

34. $5\frac{3}{6} \times 3\frac{4}{5} =$ ___

35. $5\frac{2}{8} \times 6\frac{2}{6} =$ ___

36. $7\frac{1}{4} \times 4\frac{7}{8} =$ ___

37. $9\frac{2}{3} \times 7\frac{1}{3} =$ ___

38. $6\frac{3}{4} \times 4\frac{4}{6} =$ ___

39. $5\frac{3}{8} \times 9\frac{1}{3} =$ ___

40. $8\frac{4}{5} \times 5\frac{1}{5} =$ ___

SECTION 1 ANSWERS

1. 6

2. 1/8 3. 2/15 4. 1/4 5. 5/48 6. 6/25 7. 1/10 8. 1/2 9. 2/9
10. 9/16 11. 2/25

12. 2 2/5 13. 2 1/3 14. 1 1/3 15. 3 1/3 16. 4 1/2 17. 3 3/4 18. 1 1/3 19. 1
20. 1/6 21. 3 3/4

22. 1 1/12 23. 3 21/32 24. 1 3/4 25. 11/12 26. 3 3/16 27. 6 1/6
28. 1 4/15 29. 4 4/15 30. 1 19/20 31. 5

32. 41 1/4 33. 9 3/10 34. 20 9/10 35. 33 1/4 36. 35 11/32 37. 70 8/9
38. 31 1/2 39. 50 1/6 40. 45 19/25

MULTIPLYING FRACTION - WORD PROBLEMS

Word Problems - Fractions Multiplication

Solve.

41. If $\frac{1}{3}$ is multiplied by another fraction the result is $\frac{2}{9}$. What is the value of the other fraction?

_____

42. To start getting in shape for the football team, Jackie ran $\frac{4}{5}$ miles every day. How many miles did she run in 20 days?

_____

43. Marcie buys a bag of 12 muffins. During the week she eats $\frac{2}{3}$ of the muffins. How many muffins did she eat?

_____

44. There was $\frac{2}{5}$ of a pizza left in the fridge. Sharon ate $\frac{1}{5}$ of the leftover pizza. How much of the pizza did Sharon eat?

_____

45. If you multiply $\frac{1}{6}$ by $\frac{3}{6}$, what is the result?

_____

46. What is the product of $\frac{3}{4}$ and $\frac{2}{4}$?

_____

47. Marin buys a bag of 12 muffins. During the week she eats $\frac{1}{8}$ of the muffins. How many muffins did she eat?

_____

48. What is the product of $\frac{3}{6}$ and $\frac{3}{6}$?

_____

49. To start getting in shape for the football team, Sandra ran $\frac{1}{3}$ miles every day. How many miles did she run in 20 days?

_____

50. If you multiply $\frac{2}{5}$ by $\frac{1}{5}$, what is the result?

_____

51. There was $\frac{7}{8}$ of a pizza left in the fridge. Marin ate $\frac{3}{8}$ of the leftover pizza. How much of the pizza did Marin eat?

_____

52. If $\frac{1}{4}$ is multiplied by another fraction the result is $\frac{1}{8}$. What is the value of the other fraction?

_____

53. If $\frac{1}{3}$ is multiplied by another fraction the result is $\frac{1}{9}$. What is the value of the other fraction?

_____

54. What is the product of $\frac{1}{5}$ and $\frac{2}{5}$?

_____

55. To start getting in shape for the football team, Sandra ran $\frac{3}{6}$ miles every day. How many miles did she run in 20 days?

_____

SECTION 2 ANSWERS

41. 2/3	42. 16	43. 8	44. 2/25	45. 1/12	46. 3/8	47. 1 1/2
48. 1/4	49. 6 2/3	50. 2/25	51. 21/64	52. 2/4	53. 1/3	54. 2/25
55. 10						

Dividing Fractions

The GED® Math Reasoning Test presents problems involving the dviding of fractions.

The focus of this lesson is to learn to divide fractions. The concept is used throughout basic math and algebra. Fractions are tested as a skill and also as word problems. The key is to see the connection between the skill and the word problem. Word problems become easier to solve when you can find the simple math within the text!

DIVIDING FRACTIONS - EXAMPLE

SKILL EXAMPLE

Find the quotient.

1. $6 \div \frac{3}{5} =$ _____

WORD PROBLEM EXAMPLE

Adibisi has six yards of fabric. He can make 1 skull cap with 3/5 of a yard of fabric. How many skull caps can Adibisi make with the fabric he has?

INDENTIFY THE QUESTION
First, identify that the question asks how many skull caps can Adibisi make

STATE YOUR PATH
>Adibisi has 6 yards of fabric
>Adibisi can make 1 cap with 3/5 of a yard of fabric
>6 ÷ 3/5 = 10 caps

Video Explanation

To view video explanation of example, go to: https://ispri.ng/JkD31

DIVIDING FRACTIONS - SKILL PROBLEMS

Fractions Division I

Find the quotient.

2. $\frac{5}{8} \div \frac{3}{4} =$ ___

3. $\frac{2}{3} \div \frac{2}{8} =$ ___

4. $\frac{1}{8} \div \frac{2}{3} =$ ___

5. $\frac{3}{5} \div \frac{2}{8} =$ ___

6. $\frac{1}{3} \div \frac{2}{6} =$ ___

7. $\frac{4}{6} \div \frac{3}{5} =$ ___

8. $\frac{2}{4} \div \frac{2}{6} =$ ___

9. $\frac{3}{5} \div \frac{1}{3} =$ ___

10. $\frac{4}{8} \div \frac{2}{3} =$ ___

11. $\frac{6}{8} \div \frac{1}{6} =$ ___

Fractions Division II

Find the quotient.

12. $1 \div \frac{1}{4} =$ ___

13. $9 \div \frac{1}{6} =$ ___

14. $7 \div \frac{1}{5} =$ ___

15. $7 \div \frac{4}{8} =$ ___

16. $4 \div \frac{3}{4} =$ ___

17. $2 \div \frac{5}{6} =$ ___

18. $3 \div \frac{2}{5} =$ ___

19. $8 \div \frac{2}{3} =$ ___

Fractions Division III

Find the quotient.

20. $4\frac{2}{6} \div \frac{2}{6} =$ ___

21. $1\frac{1}{3} \div \frac{2}{3} =$ ___

22. $7\frac{6}{8} \div \frac{6}{8} =$ ___

23. $7\frac{2}{3} \div \frac{2}{3} =$ ___

24. $9\frac{1}{4} \div \frac{1}{4} =$ ___

25. $4\frac{1}{5} \div \frac{4}{5} =$ ___

26. $8\frac{4}{6} \div \frac{2}{6} =$ ___

27. $5\frac{3}{4} \div \frac{3}{4} =$ ___

Fractions Division IV

Find the quotient.

28. $3\frac{3}{8} \div 3\frac{3}{8} =$ ___

29. $5\frac{1}{6} \div 4\frac{1}{6} =$ ___

30. $1\frac{2}{3} \div 5\frac{1}{3} =$ ___

31. $3\frac{4}{5} \div 1\frac{1}{5} =$ ___

32. $4\frac{5}{8} \div 5\frac{1}{8} =$ ___

33. $9\frac{1}{4} \div 2\frac{3}{4} =$ ___

34. $7\frac{4}{5} \div 1\frac{3}{5} =$ ___

35. $6\frac{1}{3} \div 6\frac{2}{3} =$ ___

SECTION 1 ANSWERS

1. 10

2. 5/6   3. 2 2/3   4. 3/16   5. 2 2/5   6. 1   7. 1 1/9   8. 1 1/2   9. 1 4/5
10. 3/4   11. 4 1/2

12. 4   13. 54   14. 35   15. 14   16. 5 1/3   17. 2 2/5   18. 7 1/2   19. 12

20. 13   21. 2   22. 10 1/3   23. 11 1/2   24. 37   25. 5 1/4   26. 26
27. 7 2/3

28. 1   29. 1 6/25   30. 5/16   31. 3 1/6   32. 37/41   33. 3 4/11   34. 4 7/8
35. 19/20

DIVIDING FRACTION - WORD PROBLEMS

Word Problems - Fractions Division

Solve.

36. Audrey bought cheese that weighs $2\frac{1}{2}$ pounds. If she divides the cheese into $\frac{2}{3}$ pound portions, how many portions will she have?

_____

37. If you divide $\frac{2}{4}$ by $\frac{2}{4}$, what is the result?

_____

38. Donald has a rope that is 12 meters long? He cuts the rope into $\frac{2}{5}$ meter lengths. How many pieces of rope did he end up with?

_____

39. Jennifer bought 6 pounds of jelly beans. If she divides the jelly beans into $\frac{1}{6}$ pound portions, how many portions can she make?

_____

40. Adam is baking mini cakes. The recipe requires $\frac{4}{5}$ cups of sugar for each cake. How many cakes can he make if he has 6 cups of sugar on hand?

_____

41. If you divide the fraction $\frac{6}{8}$ by another fraction and the result is 2, what is the value of the other fraction?

_____

42. Jake has a rope that is 12 meters long? He cuts the rope into $\frac{3}{4}$ meter lengths. How many pieces of rope did he end up with?

_____

43. Allan is baking mini cakes. The recipe requires $\frac{1}{3}$ cups of sugar for each cake. How many cakes can he make if he has 6 cups of sugar on hand?

_____

44. Jackie bought 6 pounds of jelly beans. If she divides the jelly beans into $\frac{3}{6}$ pound portions, how many portions can she make?

_____

45. Sandra bought cheese that weighs $2\frac{1}{2}$ pounds. If she divides the cheese into $\frac{1}{3}$ pound portions, how many portions will she have?

_____

46. If you divide the fraction $\frac{3}{5}$ by another fraction and the result is $1\frac{1}{2}$, what is the value of the other fraction?

_____

47. If you divide $\frac{7}{8}$ by $\frac{6}{8}$, what is the result?

_____

48. If you divide $\frac{3}{4}$ by $\frac{3}{4}$, what is the result?

_____

49. Amy bought 6 pounds of jelly beans. If she divides the jelly beans into $\frac{2}{8}$ pound portions, how many portions can she make?

_____

50. Paul has a rope that is 12 meters long? He cuts the rope into $\frac{1}{6}$ meter lengths. How many pieces of rope did he end up with?

_____

SECTION 2 ANSWERS

36. 3 3/4	37. 1	38. 30	39. 36	40. 7 1/2	41. 3/8	42. 16	43. 18
44. 12	45. 7 1/2	46. 2/5	47. 1 1/6	48. 1	49. 24	50. 72